Summary and Analysis of

THE SIXTH EXTINCTION

An Unnatural History

Based on the Book by Elizabeth Kolbert

WORTH BOOKS
SMART SUMMARIES

This Worth Books book is based on the January 2015 paperback edition of *The Sixth Extinction* by Elizabeth Kolbert, published by Picador.

ISBN: 978-1-5040-4678-7

Worth Books
180 Maiden Lane
Suite 8A
New York, NY 10038
www.worthbooks.com

Worth Books is a division of Open Road Integrated Media, Inc.

Contents

Context

Elizabeth Kolbert travels the world to meet with scientists researching the ecological impact of human activity. As a staff writer for the *New Yorker,* Kolbert first wrote about humanity's role in the extinction of Panama's golden frog in 2009.

Intended for the general reader, *The Sixth Extinction* serves as an urgent alarm about how the effects of man-made climate change and other outcomes of global trade and human activity are moving at an unstoppable pace. Kolbert synthesizes troves of research and historical anecdotes to wake up her readers to the unintentional consequences of our behaviors, from unprecedented carbon emissions and industrial agriculture to invasive species and overexploitation. Since the publication of Rachel Carson's *Silent Spring* raised awareness of the ecological damage wrought by DDT, studies continue to show that the delicate balance of life on earth is nearing a point of no return. As decades keep passing with worsening global conditions, Kolbert aims to reframe these issues through the catastrophic lens they deserve.

Overview

We are living through one of the most unique events in earth's history, significant enough to count among our planet's rarest occurrences over the past 450 million years. Interestingly, we humans are causing this event—a mass extinction, the sixth of its kind. Every species on the planet has evolved over millennia in order to thrive in its natural environment alongside a host of other species. However, human activity from the Ice Age, through the Industrial Revolution, and up to today has created a set of highly unnatural global circumstances. Drawing on the work of numerous scientists, Elizabeth Kolbert presents thirteen species that have been directly impacted by an array of unintended, yet catastrophic, events.

Humans are truly changing the shape of the world. Modern global trade and travel have allowed for flora and fauna from all corners of the world to be rearranged within the biosphere, a process resulting in a New Pangaea. Unfortunately, just as Old World diseases infected indigenous New World populations, invasive species are dominating their adopted ecosystems and destabilizing environmental equilibriums. Among the victims of this New Pangaea are Central America's golden frog, along with all the world's amphibians, which face near extinction due to an invasive species of fungus spread by global trade. Similarly, North America's little brown bats are infected with a human-spread fungus that interrupts hibernation and results in death.

As humans have dotted the globe with roads, dams, and pipelines, populations of species have been separated. This phenomenon has disrupted the local habitat of army ants in the Andes Amazon, which supports up to three hundred other species in the rainforest, setting the stage for a catastrophic domino effect. On the other side of the world, the Sumatran rhino is on the brink of extinction due to the market value of exotic woods logged from Southeast Asian forests. As loggers enter the forest and alter the natural habitat of the rhino, the animals are unable to mate and recolonize.

Most alarmingly, the planet's coral reefs and tropical rainforests are in grave danger as the burning of fossil fuels continues unabated, emitting enough carbon dioxide to drastically change the atmosphere and global temperatures. Coral reefs are especially susceptible to the increase in atmospheric carbon levels. As CO_2 is absorbed into the ocean, the pH decreases and prevents life-sustaining coral from calcifying and growing. While plants absorb CO_2 to generate photosynthesis, the accompanying temperature rise is too disruptive to the environment for most plant species to survive, especially in temperamental rainforests. These two ecological support structures—coral reefs and rainforests—harbor hundreds of thousands of species and are just as vulnerable to climate change as cold-climate species.

The Sixth Extinction focuses merely on the visible tip of the melting iceberg. Hundreds of thousands of species face extinction in the coming century, and while we are starting to wake up to the consequences of our behavior, hope for an improved situation is hard to come by. Kolbert argues the global economy must be drastically altered to stave off genuine devastation. While she doesn't anticipate such changes being made, she chooses to honor the scientists and scholars taking responsibility and trying to save species who are helpless against mankind.

Summary

Prologue

About two hundred thousand years ago, a new, not yet named species appeared on the earth in Africa. Although not particularly swift or strong, they were resourceful. Before long, they had crossed rivers, mountain ranges, and oceans without difficulty, adapting to various climates through innovation. More than one hundred thousand years passed. They traveled to faraway lands, bringing a host of germs and animals with them. The ecosystem that greeted them was forced to adapt, but often it was unable to, and other species died. Another one thousand years passed, and this species, human beings, inhabited every corner of the globe. They discovered how to use energy from deep in the ground, which alters the atmosphere and the climate. Some animals adjusted by moving—up into the mountains, into deeper water, to the poles—but thousands of species were unable to survive. No other animal has altered life on earth in the way humans have.

The planet has experienced five other periods of enormous change that resulted in mass extinctions. This is the story of the sixth. In the first part of her book, Kolbert discusses creatures that are already extinct and the history that led to our understanding of these mass extinctions. The second part of the book is concerned with the present, and looks at the catastrophic changes happening in the rainforests, the reefs, the mountains, and even our own backyards.

Chapter I: The Sixth Extinction

Kolbert dedicates her first chapter to the modern plight of the golden frog (*Atelopes zeteki*). Once abundant in Panama's rainforests, golden frogs now serve as the figurehead for the worldwide endangerment of all amphibians. In 2002, scientists, researchers, and local Panamanians noticed a drastic decline in golden frogs, and by 2004, efforts were underway to preserve the existing population in captivity and to hunt out the cause of their disappearance. Amphibians survived millions of years, dating back to before the dinosaurs, across virtually every habitat, from rainforest to desert to the Arctic Circle. However, their recent disappearance has occurred irrespective of geography and across nearly all species.

As the title suggests, there have been five mass extinctions, the first being 450 million years ago. At the center of the present sixth extinction is the exceptional fact that it is unintentionally caused by one species: us. Kolbert reveals the mysterious amphibian-killer to be a microscopic fungus called Bd, which disrupts normal functioning of the creatures' skin, resulting in the equivalent of cardiac arrest. Human activity is accountable for the global distribution of Bd, which is now so widespread across the planet that scientists believe it is impossible to repopulate the golden frog and other threatened amphibians in the wild.

Need to Know: *The Sixth Extinction* opens with the recent and rapid death of amphibians, which is caused by human activity and is occurring at a rate forty-five thousand times higher than the baseline. Once thought nearly impervious to extinction, amphibians are now the world's most endangered class of animals.

Chapter II: The Mastodon's Molars

Extinction theory dates back to the eighteenth century and French scientist Georges Cuvier, who famously studied the fossilized remains of the American mastodon. Cuvier gained renown for suggesting extinction as a widespread phenomenon, and for sensationally asserting "the existence of a world previous to ours," an idea that captivated the Age of Enlightenment and the likes of Thomas Jefferson. While not able to refine a theory of the extinction in his own lifetime, Cuvier did catalog the remains of forty-nine extinct species, including the cave bear, the giant sloth, and the pterodactyl. He also advanced the notion of catastrophism, that a cataclysmic event could cause the end of a species. As for the American mastodon, its demise resulted from humans hunting megafauna during the ice age thirteen thousand years ago.

Need to Know: Kolbert paints a picture of New World intellectuals sitting atop fossils of undiscovered species while still ignorant of their own symbiotic relationship to the natural world. Once Cuvier arrived at the turn of the nineteenth century, the "previous" world of extinct fauna gained much interest and scrutiny.

Chapter III: The Original Penguin

Cuvier's theory of extinction became overshadowed by its great scientific cousin, Charles Darwin's theory of evolution. Darwin and his contemporary, geologist Charles Lyell, criticized Cuvier and other catastrophists by asserting the primary cause of extinction to be natural selection, the same gradual mechanism behind evolution—not catastrophic events. Kolbert recounts a notable extinction that occurred within Darwin's own lifetime. The great auk was a three-foot-tall, flightless bird that inhabited rocky shores throughout the North Atlantic. Its population numbered in the millions until contact with Scandinavian settlers in the tenth century. The most notable extermination of the great auk occurred near Newfoundland, where numerous expeditions exploited a colony of one hundred thousand auks in the span of two centuries, initially to feed malnourished settlers and eventually for the lucrative feather trade. This overexploitation led to extinction in 1844, when the last known pair of auks were haphazardly strangled and their last egg was cracked by three men off the coast of Iceland.

Need to Know: Kolbert traces the prominence of extinction in scientific literature alongside the study of evolution and its primary engine, natural selection. While the auk and the dodo were killed off by humans due to overexploitation, prominent scientists of the day could not recognize mankind's connection to extinction.

Chapter IV: The Luck of the Ammonites

While studying plate tectonics in the late 1970s, renowned geologist Walter Alvarez and his father, Nobel laureate and physicist Luis Alvarez, discovered a thin layer of iridium, an element found abundantly in asteroids. This iridium layer was found in various sites around the world and served as a geological marker for an abrupt global event. The Alvarezes suggested this marker corresponded with the mass extinction of the dinosaurs sixty-six million years ago, wherein vaporized iridium dust created by an asteroid's impact incinerated nearly everything on the earth's surface in a matter of minutes.

Kolbert joins modern geologists in New Jersey to study ammonites, a class of marine creatures that did not survive the great asteroid's impact, along with three-quarters of all species. Like the great auks, which could not outrun humans, ammonites and countless other species on the planet endured millennia of natural selection but could not escape the asteroid.

Need to Know: Our current understanding of the end of the Cretaceous period and the previous mass extinction is a recent discovery. This shift in thinking among the scientific community solidified that evolution through natural selection may equip flora and fauna to thrive, yet one dramatic change to the environment can cause widespread extinction.

Chapter V: Welcome to the Anthropocene

Mass extinctions began more than 450 million years ago, when fossil records indicate severe changes in climate occurred. These catastrophic changes were caused by the sharp rise and fall of atmospheric CO_2 levels and altered the course of life on our planet. Too much carbon dioxide leads to increased temperatures and rising sea levels, and when carbon dioxide levels drop, temperatures fall and oceans are absorbed in ice, a process called glaciation. The first mass extinction, the end-Ordovician, was a period with such extreme glaciation that ninety percent of life on earth did not survive.

Humans have dominated the natural processes of the planet, altering nearly half of all dry land, damming or diverting most major rivers, using more than half of the world's fresh water runoff, and changing the atmosphere through deforestation and fossil fuel combustion. The past two centuries have shown a forty percent increase in atmospheric CO_2 levels and a doubling of the concentration of methane, an even more potent greenhouse gas.

Need to Know: Aside from the asteroid event that ended the Cretaceous period, the previous four mass extinctions have been tied to global cooling and warming events that triggered the end of most life on earth. Today, human activity has caused similarly drastic changes in atmospheric conditions.

Chapter VI: The Sea Around Us

Since the beginning of the Industrial Revolution, the burning of fossil fuels (coal, oil, and natural gas) has added approximately 365 billion metric tons of carbon dioxide to the atmosphere, and deforestation has contributed an additional 180 billion metric tons. As this process continues, the total concentration of carbon dioxide has reached the highest levels in the last eighty million years. In nature, where ocean meets air, gases from the atmosphere are absorbed and dissolved. When carbon dioxide dissolves in water, it forms carbolic acid, which lowers the pH level, resulting in a precipitous decline in the pH of the planet's oceans. Many marine species rely on a delicate balance in pH for calcification of shells and coral, so they are highly susceptible to ocean acidification. On a global scale, based on current rates of CO_2 release, marine biologists predict entire marine ecosystems are on pace to "crash" due to ocean acidification.

Need to Know: Kolbert outlines the process of ocean acidification, which results from the rapid increase in atmospheric carbon dioxide levels. Human civilization has produced a historic level of carbon emissions which threaten nearly all marine life.

Chapter VII: Dropping Acid

Kolbert visits Australia's Great Barrier Reef, where coral cover has declined by fifty percent in the past thirty years. Coral reefs—part animal, part vegetable, part mineral—are a unique ecological structure responsible for providing food and protection to hundreds of thousands of marine species. Their construction and growth requires aragonite, a crystal form of calcium carbonate, which disintegrates under acidic conditions. Therefore, as carbon emissions persist, lowered pH inhibits the natural regeneration of coral. Rising ocean temperatures also disrupt equilibrium between coral and other species within the reef, leading to "coral bleaching" and eventual death of coral colonies. Scientists predict "reefs will be the first major ecosystem in the modern era to become ecologically extinct," likely in the next fifty years.

Need to Know: Coral reefs harbor half a million other species, yet they are highly susceptible to ocean acidification caused by carbon emissions. Large swaths of coral have already been lost around the world and may soon disappear, eliminating a vital ecosystem from the world's oceans.

Chapter VIII: The Forest and the Trees

The ecological diversity of the tropics is much greater than toward the earth's poles, a relationship known as the latitudinal diversity gradient. This diversity mimics coral reefs, where trees are the primary organisms supporting the life of forests. Plants themselves rely on other organisms for survival—birds disperse seeds and combat harmful insects, and helpful insects pollinate. This relationship, called the species-area relationship, has been studied by scientists to predict the extinction risk of global warming in regions like the Andean Amazon. Since tropical plant species are more sensitive to altitude and changes in temperature caused by global warming, they suffer when this delicate balance collapses. While some plant species thrive from increased carbon dioxide levels, their general lack of mobility threatens their survival amid increased temperatures. It is predicted nearly one-fourth of all tropical plant species will be fated to extinction in less than forty years. Beyond that, it is impossible to predict how plants, ill equipped to adapt to sudden change, will be able to survive.

Need to Know: Global warming threatens cold-loving species as much as tropical species. The vast and delicate ecological diversity of the tropics is just as fragile; 24% of all species will become extinct by 2050 due to global warming and climate change. Prominent scientists have called the scenario "apocalyptic."

Chapter IX: Islands on Dry Land

Humans have directly transformed more than half of the planet's fifty million square miles of ice-free land, and the remaining, untouched land has been separated by pipelines, ranches, hydroelectric projects, and more. Kolbert visits a group of scientists who study the effect of this fragmentation in the Brazilian Amazon by intentionally isolating plots of land into twenty-five acre "islands" of pristine rainforest surrounded by cleared scrub. Fragmentation separates species into smaller populations unable to support a stable number of members who can insulate themselves against natural calamities. Subsequently, species members outside of the islands cannot reach the dwindling survivors to recolonize. This phenomenon exacerbates extinction. Kolbert illustrates how one species of army ants lives in association with hundreds of other species. This level of interdependence has led scientists to estimate that the planet's tropical rainforests lose nearly five thousand species every year due to fragmentation.

Need to Know: As humans fragment the natural environment, small populations of isolated species become especially vulnerable.

Chapter X: The New Pangaea

Kolbert joins scientists in the northeastern United States to study the dwindling population of little brown bats, which have fallen victim to white-nose syndrome, caused by an invasive species of fungus accidentally imported from Europe by humans. While native species have always been threatened by foreign species—from fungus and bacteria to plants and rodents—modern travel and global trade have accelerated this process by accidentally transporting flora and fauna around the world. In effect, humans have removed important geographic boundaries to create a "New Pangaea." As various species are redistributed into non-native territories, they often thrive at the expense of local populations who are unable to adapt. Victims of this phenomenon range worldwide and include the American chestnut, the Central American wolfsnail, and the Guam flycatcher. From

2007 until 2013, it is estimated that white-nose syndrome has spread to twenty-two states and five Canadian provinces, killing more than six million bats. In less than ten years, the little brown bat has become an endangered species.

Need to Know: While humans have fragmented land and isolated species, they have also introduced invasive species as a potent mechanism of the sixth extinction. The increased pace and volume of global trade has precipitated the extinction and steep decline of numerous species, such as the little brown bat.

Chapter XI: The Rhino Gets an Ultrasound

Of all the endangered species humans work to save, large animals are the most visible and most vulnerable. Kolbert takes readers to the Cincinnati Zoo, where zoologists attempt to breed the Sumatran rhino, whose ancestors were victims of habitat fragmentation in the rainforests of Southeast Asia during the last century. Humans have been a major threat to populations of megafauna like rhinos and the moa of New Zealand since well before the industrial age. These outsized creatures once dominated the food chain, but ten to fifteen thousand years ago, as humans began to hunt large animals for food, "'the rules of the survival game' changed." Many large animals' primary evolutionary weakness in the face of human civilization is a long gestation period, which prevents repopulation at the same rate of decline.

Need to Know: With fewer than one hundred members worldwide, the Sumatran rhino is another example of the threat posed by fragmentation, and efforts to save the remaining population outside of its natural habitat are proving futile. While prevailing theories of the Anthropocene suggest the impact of humans began with the Industrial Revolution, our species has threatened the populations of megafauna like the rhino since the Stone Age.

Chapter XII: The Madness Gene

For nearly one hundred thousand years, Neanderthals ranged from Europe to the Middle East and shared many traits with humans: constructing tools, building shelters, fashioning clothing, and caring for members of their tribe. Thirty thousand years ago, Neanderthals disappeared. Archaeological records indicate humans made contact with Neanderthals in various regions, and molecular sequencing projects show the genetic code of modern humans is one to four percent Neanderthal, evidence the two species interbred and cohabitated. Paleogenticists like Svante Pääbo study why humans overtook Neanderthals to become the dominant species. One theory is humans' desire to disperse and explore, leading our species to venture across land bridges and open water to places like Australia and the Americas. Pääbo and others investigate our genetic code to isolate this behavior, without which Neanderthals and many other species would still exist.

Need to Know: Neanderthals and apes share many traits with humans, including socialization and the use of tools, yet humans have spread across the globe and overtaken many species. Some key differences between humans and our genetic ancestors are behaviors like exploration and pursuit of art, which suggest humans have a genetic desire to change our worlds.

Chapter XIII: The Thing with Feathers

Concerned citizens continue to raise awareness of threatened species, which has led to landmark protections around the world, like the Endangered Species Act of 1974, which was spurred by Rachel Carson's *Silent Spring*. Within the scientific community, efforts are underway to freeze and preserve the genetic information of all species on the planet to ensure absolute extinction will not happen. The arrival and expansion of the human species has led to another extinction event, as unique as an asteroid impact at the end of the Cretaceous period.

Need to Know: Kolbert issues a call to her readers to take responsibility for our legacy by embracing our connection to the natural world. The fate of virtually all species, including our own, is at stake.

Timeline

460,000,000 BCE (approx.): Earliest land plants being to populate the land midway through the Ordovician period.

450,000,000 BCE (approx.): End-Ordovician (first) Extinction, caused by glaciation. Graptolites become extinct.

370,000,000 BCE (approx.): Late Devonian (second) Extinction, caused by anoxia.

315,000,000 BCE (approx.): Earliest reptiles.

252,000,000 BCE: End-Permian (third) Extinction, caused by global warming, ocean acidification, and increase in atmospheric sulfides.

200,000,000 BCE (approx.): Late Triassic (fourth) Extinction, caused by volcanic activity.

148,000,000 BCE (approx.): Earliest birds.

130,000,000 BCE: Earliest flowering plants.

65,000,000 BCE: End-Cretaceous (fifth) Extinction, caused by an asteroid impacting the earth in the Gulf of Mexico. Dinosaurs and ammonites become extinct.

55,000,000 BCE (approx.): Earliest primates appear.

35,000,000 BCE (approx.): Glaciation of Antarctica.

15,000,000 BCE (approx.): Earliest Great Apes appear.

2,580,000 BCE (approx.): Ice ages begin.

300,000 BCE: Neanderthals emerge.

200,000 BCE: *Homo sapiens* emerge.

40,000 BCE: *Homo sapiens* make contact with Neanderthals; Neanderthals become extinct shortly after.

11,700 BCE (approx.): Last ice age ends. Megafauna extinction event. American mastodon becomes extinct.

1739: French soldiers discover American mastodon fossils in Big Bone Lick, Kentucky, and ship them down the Mississippi River and across the Atlantic to Paris for study.

1769: James Watt patents the steam engine, and the Industrial Revolution begins to take root in Great Britain.

April 4, 1796: Georges Cuvier presents his lecture built upon the fossils from Big Bone Lick that asserts "the existence of a world previous to ours."

1806: Cuvier publishes the name of a unique, extinct species, the American mastodon, in a Parisian newspaper.

1844: Great auk becomes extinct.

November 24, 1859: Charles Darwin publishes *On the Origin of Species*, the foundational text of evolutionary biology.

June 1980: George Alvarez presents evidence of the iridium layer and his theory of the asteroid that ended the Cretaceous period.

1993: Bd outbreak begins in Queensland, Australia.

February 2006: Fungus linked to white-nose syndrome is first identified on little brown bats in New York State.

2014: Stuart Pimm's article is published in *Science* magazine; it outlines how current extinction rates have risen to about one thousand times the likely background rate of extinction.

Cast of Characters

Walter Alvarez: Modern earth science researcher best known for his theory that the Cretaceous period ended with an asteroid impact, which killed nearly all life on earth in a matter of minutes.

Ken Caldeira: Ecologist, atmospheric scientist, and author of the paper "Anthropogenic carbon and ocean pH: The coming centuries may see more ocean acidification than the past 300 million years."

Georges Cuvier: Nineteenth-century French naturalist and zoologist, and "the father of paleontology"; Cuvier proved to be an important scientific figure during his lifetime and his cataloging of forty-nine extinct species, including the American mastodon, laid the foundation of early extinction theory, which included the theory of catastrophism. Interestingly, Cuvier did not believe in the natural selection theories of evolution.

Charles Darwin: Nineteenth-century English natural historian and geologist who is best known for his landmark contributions to evolution science, especially natural selection, a process he believed applied to the origin of species as well as their extinction.

Edgardo Griffith: Native Panamanian and director of El Valle Amphibian Conservation Center (EVACC) who devoted nearly all of his adult life to local amphibians.

Al Hicks: Supervisor for New York State's Department of Environmental Conservation who has surveyed the decimation of the little brown bat as a result of white-nose syndrome.

Thomas Jefferson: Third president of the United States and principal author of the Declaration of Independence. During his lifetime, Jefferson had a keen interest in fossil remains and weighed in on Georges Cuvier's work with the American mastodon fossils.

Neil Landman: Staff member at American Museum of Natural History and paleontologist with a specialty in ammonites and cephalopods; he took Kolbert on a field trip to a region rich in ammonites in suburban New Jersey.

Charles Lyell: Nineteenth-century geologist whose theories challenged Georges Cuvier's outlook on extinction. Lyell greatly aided Charles Darwin and Alfred Russel Wallace in their work on natural selection and evolution.

Alfred Newton: Nineteenth-century English ornithologist who corresponded with Charles Darwin and became the first advocate for the preservation of seabirds due to overexploitation.

Svante Pääbo: Swedish evolutionary geneticist and anthropologist responsible for sequencing the Neanderthal genome and identifying a new hominid species, the Denisovans, from a finger bone fragment found in Siberia.

Stuart Pimm: British-born American ecologist whose work in biodiversity and conservation biology has made large contributions to Anthropocene extinction literature.

Suci: A female Sumatran rhino, was one of fewer than one hundred surviving members of her species, and fewer than ten in captivity worldwide. Suci was part of the Cincinnati Zoo's breeding program up until her death in March of 2014.

David Wake: A University of California-Berkeley emeritus professor; former director and curator of herpetology of the Museum of Vertebrate Zoology at UC Berkeley; author of "Are We in the Midst of the Sixth Mass Extinction? A View from the World of Amphibians," which was published in August 2008 in *Proceedings of the National Academy of Sciences* and was the article responsible for the primary definition of mass extinction used by Kolbert.

E. O. Wilson: A modern American biologist, conservationist, and leading expert of ants and their relationship to landscape fragmentation; he has concluded the extinction rate of tropical insects is "on the order of 10,000 times greater than the naturally occurring background rate."

Jan Zalasiewicz: British author and lecturer specializing in geology and the study of graptolites, which went extinct during the end-Ordovician period. His research has been applied to the study of the present Anthropocene extinction event.

Direct Quotes and Analysis

"Those of us alive today not only are witnessing one of the rarest events in life's history, we are also causing it."

This quote arrives early in the text and introduces the point of view Kolbert takes throughout, an approach that shows the reader his or her connection to historical events. Kolbert is referring to the current mass extinction—the sixth in earth's history. In her book, she lays out the cause for the mass extinction: humanity. By destroying animals' habitats through farming and building cities, and by expelling greenhouse gases that change the temperature of the climate, human life is directly causing the sixth extinction.

"A vast cloud of searing vapor and debris raced over the continent, expanding as it moved and incinerating anything in its path. 'Basically, if you were a triceratops in Alberta, you had about two minutes before you got vaporized' is how one geologist put it to me."

Enormous change can happen in the blink of an eye, such as when a giant meteor's crash to earth caused the extinction of the dinosaurs. Geologically speaking, this is happening in today's landscape because of human civilization. By destroying animals' territories and altering the planet's atmosphere, we are creating an inhospitable environment that is already leading to mass extinctions around the world, from the Arctic to the jungles to the oceans.

"The reason this book is being written by a hairy biped, rather than a scaly one, has more to do with dinosaurian misfortune than with any particular mammalian virtue."

This terse quote speaks to science uncovering random catastrophes that completely disrupt the "divine" order of our planet. Life on earth isn't simply a matter of evolution—the weak die out, the strong get stronger—because random catastrophic events disrupt the normal order of things. The dinosaurs didn't die because they weren't strong enough; they died because the asteroid killed them all. The Great Auk didn't disappear because they weren't cunning enough; they are gone because humans hunted them to extinction.

"One of the defining features of the Anthropocene is that the world is changing in ways that compel species to move, and another is that it's changing in ways that create barriers—roads, clear-cuts, cities—that prevent them from doing so."

In order to explain the significance of fragmentation, Kolbert demonstrates two sides of this one cause of extinction and discusses why it is so devastating and ironic. When a climate gets warmer, animals are forced to move to cooler locales to survive. Unfortunately, humans have fragmented the planet with their dams, cities, and roads, preventing the animals from moving—and so they perish.

"Though it might be nice to imagine there once was a time when man lived in harmony with nature, it's not clear that he ever really did."

Although in recent years humans have been attempting to counter the effects of the Anthropocene by trying to save endangered species from extinction, this impulse stands in stark contrast to our species' historical record. It is only lately that we've truly understood the impact of our actions on the natural habitats and populations of other species. Being so late to this comprehension opens up efforts like zoo breeding programs to accusations of futility.

"The Neanderthals lived in Europe for more than a hundred thousand years and during that period they had no more impact on their surroundings than any other large vertebrate. There is every reason to believe that if humans had not arrived on the scene, the Neanderthals would be there still, along with the wild horses and the woolly rhinos."

Humans are not particularly fast, strong, large, or tough. Yet what we lack in athleticism, we make up in resourcefulness, creativity, and constant innovation—the exact qualities that have led to our planet's current crisis. Our ability to imagine, create, and build set us apart from every other species on earth, including our close cousins the Neanderthals. Our capacity to change the planet—the key to helping us flourish in all types of climates—is ours alone. For the most part, other plant and animal species have been unable to adapt to the changing climate at the rate that we are altering it.

Trivia

1. Due to the chytrid fungus Bd, amphibians are in critical danger and are considered to be the most endangered class of animals, with an extinction rate as much as forty-five thousand times higher than the background extinction rate.

2. CO_2 levels in the atmosphere have now surpassed three hundred parts per million, the highest concentration in eight hundred thousand years.

3. The asteroid theory of extinction taught to schoolchildren around the world—which ended the Cretaceous period 66 million years ago and

killed off the dinosaurs—was only popularized quite recently. It is based upon the June 1980 publication of "Extraterrestrial Cause for Cretaceous-Tertiary Extinction" written by Luis W. Alvarez, Walter Alvarez, Frank Asaro, and Helen V. Michel.

4. Once a year, right after the full moon, the corals of the Great Barrier Reef engage in a mass spawning—or massive synchronized sex—by releasing bundles of sperm and eggs into the water simultaneously. This forsters genetic diversity.

5. The state of the Great Barrier Reef has declined by 50% in the past thirty years. Coral cover in the Caribbean Sea has declined by nearly 80%.

6. Worldwide research indicates that the tropics are home to as many as thirty million species of arthropods.

7. Estimates indicate up to five thousand individual insect species go extinct every year.

8. President Barack Obama listed *The Sixth Extinction* as one of his books for summer vacation reading in 2015.

9. It's believed that modern humans led to the Neanderthals' extinction. But before the Neanderthals died out, humans reproduced with them. Genetically, most people are up to 4% Neanderthal.

10. When bats hibernate in the winter, their body temperatures drop fifty to sixty degrees (close to freezing), their immune systems shut down, and their heartbeats slow. Kolbert writes that they "fall into a state close to suspended animation."

What's That Word?

Ammonite: A type of shelled marine mollusk that died out during the Cretaceous-Paleogene extinction event.

Anoxia: Low oxygen conditions.

Background extinction rate: The natural rate of species extinctions that would occur without the impact of human activity.

Catastrophism: The theory, popularized by Georges Cuvier in the nineteenth century, that changes to the earth's geology have resulted from sudden, disruptive, and unnatural events.

Glaciation: The process, condition, or result of being covered by glaciers; Periods of glaciation are often caused by a decrease in atmospheric carbon dioxide.

Graptolite: A once vast and extremely diverse class of marine organisms that thrived during the Ordovician period and disappeared during the first mass extinction, about 444 million years ago.

Invasive species: Plants, animals, or pathogens that are alien to an ecosystem and harmful to other species.

Landscape fragmentation: Unnatural division of continuous habitats into a number of smaller regions.

Mass extinction: Events that "eliminate a significant proportion of the world's biota in a geologically insignificant amount of time."

Mastodon: A large, extinct, elephant-like mammal of the Miocene to Pleistocene epochs; it had teeth of a relatively primitive form and number.

Megafauna: Land mammals of a particular region, habitat, or geological period.

Neanderthal: An extinct species of human with a receding forehead and prominent brow ridges that was widely populated in ice age Europe approximately 120,000–32,000 years ago.

New Pangaea: The modern process of interchanging species across biospheres as a result of global trade and travel; also refers to the ancient landmass that was home to all terrestrial species until plate tectonics separated the supercontinent.

Ocean acidification: The ongoing decrease in the pH of the earth's oceans, caused by the increase of carbon dioxide from the atmosphere.

Overexploitation: Harvesting of species from the wild at rates faster than natural populations can recover.

White-nose syndrome: A fungus-based, emergent disease affecting hibernating bats that has spread across North America, killing millions of bats.

Critical Response

- An Andrew Carnegie Medal for Excellence nominee for nonfiction
- A Kirkus Prize finalist for nonfiction
- A *Library Journal* Top Ten Best Book
- A *Los Angeles Times* Book Prize for Science & Technology winner
- A National Book Critics Circle Award finalist
- A *New York Times* bestseller
- A *New York Times Book Review* 10 Best Book of 2014
- A PEN/E.O. Wilson Literary Science Writing Award nominee
- A Pulitzer Prize for General Nonfiction winner

"Arresting . . . Ms. Kolbert shows in these pages that she can write with elegiac poetry about the vanishing creatures of this planet, but the real power of her book resides in the hard science and historical context she delivers here, documenting the mounting losses that human beings are leaving in their wake."

—*The New York Times*

"[*The Sixth Extinction*] is a wonderful book, and it makes very clear that big, abrupt changes can happen; they're not outside the realm of possibility. They have happened before, they can happen again."

—President Barack Obama

"The factoids Kolbert tosses off about nature's incredible variety—a frog that carries eggs in its stomach and gives birth through its mouth, a wood stork that cools off by defecating on its own legs—makes it heartbreakingly clear, without any heavy-handed sermonizing from the author, just how much we lose when an animal goes extinct." —*Bookforum*

"Throughout her extensive and passionately collected research, Kolbert offers a highly readable, enlightening report on the global and historical impact of humans. . . . A highly significant eye-opener rich in facts and enjoyment." —*Kirkus Reviews* (starred review)

About Elizabeth Kolbert

Elizabeth Kolbert was born in the Bronx, New York, in 1961 but grew up in Larchmont, New York, and attended Mamaroneck High School. Upon graduation in 1979, Kolbert studied literature at Yale University and then pursued further study at Universität Hamburg in Germany as a Fulbright Scholar.

Kolbert's journalism career began in 1983 when she started working with the German syndicate of the *New York Times* on a freelance basis; she later returned to the United States and the paper's Albany, New York, offices. In 1998, she began a staff-writing job at the *New Yorker*, where her cultural and environmental writing began in earnest.

Kolbert's work has been celebrated across journalism and science circles for over a decade, culminating with the 2015 Pulitzer Prize for General Nonfiction for *The Sixth Extinction*. As of 2016, she teaches at Williams College in Williamstown, Massachusetts, where she lives with her husband and three children.

For Your Information

Online

"The 100 Best Nonfiction Books: No 1—The Sixth Extinction by Elizabeth Kolbert (2014)." TheGuardian.com

"Elizabeth Kolbert: 'The Sixth Extinction: An Unnatural History' | Talks at Google." YouTube.com

"In The World's 'Sixth Extinction,' Are Humans The Asteroid?" NPR.org

"Nature's Bounty: 5 Earth Day Reads." EarlyBirdBooks.com

"The Sixth Extinction: Earth is on the brink of another massive loss of animal species but this time the calamity isn't an asteroid or ice age . . ." Independent.co.uk

Books

Crossing Open Road by Barry Lopez
Field Notes from a Catastrophe: Man, Nature, and Climate Change by Elizabeth Kolbert
Guns, Germs, and Steel: The Fate of Human Societies by Jared Diamond
Sapiens: A Brief History of Humankind by Yuval Noah Harari
The Sea Around Us by Rachel Carson
Silent Spring by Rachel Carson
The Solace of Open Spaces by Gretel Ehrlich
This Changes Everything: Capitalism vs. The Climate by Naomi Klein

Bibliography

"Elizabeth Kolbert: 'The Sixth Extinction: An Unnatural History' | Talks at Google." YouTube. March 13, 2014. https://www.youtube.com/watch?v=x00LP0QfRTk&feature.

Kolbert, Elizabeth. *The Sixth Extinction: An Unnatural History*. New York: Picador, 2014.

"The Sixth Extinction." Macmillan. Accessed October 30, 2016. http://us.macmillan.com/thesixthextinction-1/elizabethkolbert.

MORE SMART SUMMARIES

FROM WORTH BOOKS

POPULAR SCIENCE

MORE SMART SUMMARIES

FROM WORTH BOOKS

HISTORY

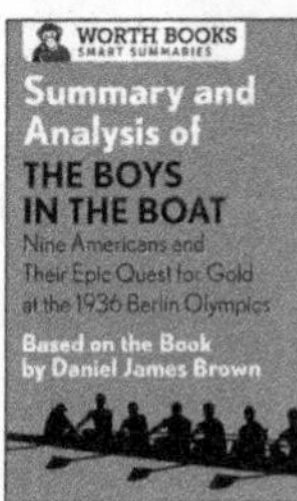

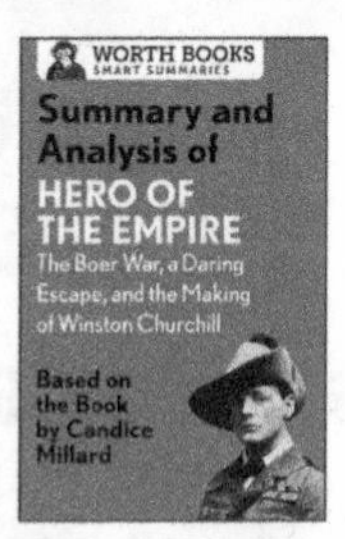

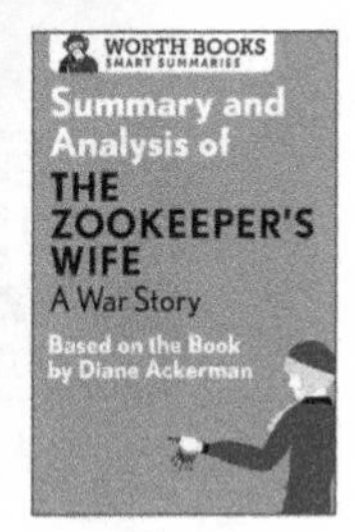

CPSIA information can be obtained
at www.ICGtesting.com
Printed in the USA
LVHW111044010422
714850LV00013B/60